Julio César Canul Ek
Antonio Alberto Vela Ávila

NUMERICAL METHODS

Julio César Canul Ek
Antonio Alberto Vela Ávila

NUMERICAL METHODS

Anthology

ScienciaScripts

Imprint

Cover image: www.ingimage.com

This book is a translation from the original published under ISBN 978-620-2-24951-5.

Publisher:
Sciencia Scripts
is a trademark of
Dodo Books Indian Ocean Ltd. and OmniScriptum S.R.L publishing group

120 High Road, East Finchley, London, N2 9ED, United Kingdom
Str. Armeneasca 28/1, office 1, Chisinau MD-2012, Republic of Moldova, Europe
Printed at: see last page
ISBN: 978-620-8-33048-4

Content

INTRODUCTION

This anthology presents a selection of fundamental concepts and methods in Numerical Methods, essential for students at the **Tecnológico Nacional de México**. Although it does not cover all topics, it seeks to illustrate how each method offers advantages and disadvantages to achieve accurate results.

Numerical methods are crucial tools for solving complex problems, especially when the functions involved have a higher level of complexity than those studied in high school. In this context, it is essential to resort to techniques that provide approximate results, as accurate as possible, in order to make informed decisions.

To facilitate the understanding and application of these methods, technological tools such as Microsoft Excel and GeoGebra will be used.

These softwares allow:

- Graphical visualisation of functions and numerical methods
- Perform accurate calculations and simulations
- Analyse and compare results
- Develop practical problem-solving skills

In the following sections, the following numerical methods will be presented:

- **Bisection method**: find roots by dividing intervals.
- **False Rule**: combines bisection with linear interpolation.
- **Fixed Point**: finds roots by successive iterations.
- **Newton Raphson**: uses the derivative to find roots quickly.
- **Secant**: uses two points to estimate the root.
- **Multiple Root Method**: solves equations with repeated roots.
- **Graphical method**: visualise the function to find roots.

These methods will enable students to understand the importance of Numerical Methods in solving complex problems and to develop the skills to apply them in various areas of science and engineering.

The integration of Excel and GeoGebra in this text will provide a more interactive and effective learning experience, enabling students:

- Explore and analyse numerical methods in a visual and practical way.
- Develop skills in the use of technological tools for problem solving.
- Apply the concepts learnt in real and practical contexts.

This anthology presents a selection of the fundamental concepts and methods that students of the **Tecnológico Nacional de México** in the subject of Numerical Methods must master.

Although it does not cover all topics, it seeks to illustrate how each method, whether open or closed interval, offers advantages and disadvantages that can lead to the same result. It is important to note that the functions discussed in this text have a higher level of complexity than those studied in the baccalaureate, so it is necessary to resort to numerical techniques that provide approximate results, as accurate as possible, in order to make informed decisions.

The methods described in the following sections will show various ways of obtaining approximate values of the roots of a function, including estimation of the percentage error. This will enable students to understand the importance of Numerical Methods in solving complex problems.

We hope that this anthology will be a valuable tool for students at the **Tecnológico Nacional de México** and for anyone interested in Numerical Methods.

Chapter 1

Closed interval methods

1.1 Bisection Method
1.2 Graphical method
1.3 False rule method

1.1 BISECTION METHOD

Considered as a Closed Interval Method, this method determines values of x that satisfy a function f(x) = 0, where, in most cases f(x) is known, real, one variable and continuous; once or twice differentiable in the region of interest, it may be a polynomial.

In this topic we give a real function of a real variable x, examine its behaviour for a very large magnitude of x and thus locate all its roots within one of its finite intervals.

One of the proposed intervals is not part of the solution, the other is solved to describe the method, while the last one is left as a proposal to be practised by the interested reader.

Given the function f(x) = $2x^2 + 5x + 1$

a) Determine in which of the following intervals [1,2], [-3,-2] and [-1,0] is the value of the variable **X** such that the graph of the function f(x) intersects the X-axis.
b) What is the value of **X** at the point where the graph crosses the X-axis.

Procedure to determine in which interval the graph crosses.

- ***For the interval [1,2], considering the interval [a, b], where a = 1 and b = 2***
- Evaluate f(a) and f(b):

$$f(a) = f(1) = 2(1)^2 + 5(1) + 1 = 8$$
$$f(b) = f(2) = 2(2)^2 + 5(2) + 1 = 19$$

- If f(a)*f(b) < 0, then the graph passes in the studied interval through the X-axis. x-axis, otherwise it does not cut the axis. In this case:

$$f(a) * f(b) = 152, es\ mayor\ que\ CERO.$$

Therefore, in the interval [1,2] it does not cross the graph of the function on the X axis and is not the object of study as it does not form part of the function.

- ***For the interval [-3,-2], considering the interval [a, b], where a = -3 and b = -2***
- Evaluate f(a) and f(b):

$$f(a) = f(-3) = 2(-3)^2 + 5(-3) + 1 = 4$$
$$f(b) = f(-2) = 2(-2)^2 + 5(-2) + 1 = -1$$

- Since f(a)*f(b) < 0, then, the graph passes through the studied interval crossing the X-axis. In this case:
-

$$f(a) * f(b) = -4, es\ menor\ que\ CERO.$$

Therefore, in the interval [-3,-2] it does cross the graph of the function on the X axis and is the object of study as it forms part of the function.

- ***For the interval [-1,0], considering the interval [a, b], where a = -1 and b = 0***
- Evaluate f(a) and f(b):

$$f(a) = f(-1) = 2(-1)^2 + 5(-1) + 1 = -2$$

$$f(b) = f(0) = 2(0)^2 + 5(0) + 1 = +1$$

- Since f(a)*f(b) < 0, then, the graph passes through the studied interval crossing the X-axis. In this case:

$$f(a) * f(b) = -2, es\ menor\ que\ CERO.$$

Therefore, in the interval [-1,0] it does cross the graph of the function on the X axis and is the object of study as it forms part of the function.

Analysis process of each interval by iterations seeking to calculate the value of X as close as possible, where X = Xi, when i = 1, 2, 3 ... n

First interval to be studied [-3, -2].

Determine the mean value that exists between [-3, -2] using the following formula:

$$x_i = a + (b - a)/2 \text{(Formula No. 1) then:}$$

Iteration 1.

Where: a = -3 and b = -2

$$x_1 = -3 + \frac{[-2 - (-3)]}{2} = -3 + 0.50 = -2.50$$

It can be seen that the first value of x_1 lies between the interval [-3, -2].

In order to start our comparison we evaluate the x-value$_1$ in the function:

$$f(x_1) = f(-2.5) = 2(-2.5)^2 + 5(-2.5) + 1 = +1$$

Interval Analysis:

$$\left[\frac{-3}{f(a) = +} \quad \frac{-2.5}{f(x_1) = +} \quad \frac{-2}{f(b) = -}\right]$$

$f(a) * f(x_1) > 0$, in the Interval [-3, -2.5], does not cut the graph to the X-axis, therefore, it is not the object of study.

$f(x_1) * f(b) < 0$, in the Interval [-2.5, -2], if it cuts the graph to the X-axis, therefore, if it is the object of study.

Iteration 2.

In the interval [-2.5, -2], where a = -2.5 and b = -2

Consider the expression again: $x_i = a + (b - a)/2$

$$x_2 = -2.5 + \frac{[-2 - (-2.5)]}{2} = -2.5 + 0.25 = -2.25$$

Again we evaluate the function, but now with the value x_2

$$f(x_2) = f(-2.25) = 2(-2.25)^2 + 5(-2.25) + 1 = -0.125$$

Interval Analysis:

$$\left[\frac{-2.5}{f(a) = +} \quad \frac{-2.25}{f(x_2) = -} \quad \frac{-2}{f(b) = -}\right]$$

$f(a) * f(x_2) < 0$, in the Interval [-2.5, -2.25], if it cuts the graph to the X-axis, therefore, if it is the object of study.

$f(x_2) * f(b) > 0$, in the Interval [-2.25, -2], does not cut the graph to the X-axis, therefore, it is not the object of study.

Percentage error up to Iteration 2.

$$E_p = \left|\frac{V_{act} - V_{ant}}{Vact}\right| * 100 \quad Fórmula\ No. 2$$

Where: V_{act} is Present value
V_{ant} en Previous value
Ep en Percentage error (%)

$$E_p = \left|\frac{-2.25 - (-2.5)}{-2.25}\right| * 100 = 11.11\%$$

Iteration 3.

In the interval [-2.5, -2.25], where a = -2.5 and b = -2.25

Consider the expression again: $x_i = a + (b - a)/2$

$$x_3 = -2.5 + \frac{[-2.25 - (-2.5)]}{2} = -2.5 + 0.125 = -2.375$$

$$f(x_3) = f(-2.375) = 2(-2.375)^2 + 5(-2.375) + 1 = +0.40625$$

Interval Analysis:

$$\left[\frac{-2.5}{f(a)=+} \quad \frac{-2.375}{f(x_3)=+} \quad \frac{-2.25}{f(b)=-}\right]$$

$f(a)*f(x_3)>0$, in the Interval [-2.5, -2.375], does not cut the graph to the X-axis, therefore, it is not the object of study.

$f(x_3)*f(b)<0$, in the Interval [-2.375, -2.25], if it cuts the graph to the X-axis, therefore, if it is the object of study.

Percentage error up to Iteration 3.

$$E_p=\left|\frac{V_{act}-V_{ant}}{Vact}\right|*100$$

$$E_p=\left|\frac{-2.375-(-2.25)}{-2.375}\right|*100=5.26\%$$

Iteration 4.

On the interval [-2.375, -2.25], where a = -2.375 and b = -2.25
Consider the expression again: $x_i=a+(b-a)/2$

$$x_4=-2.375+\frac{[-2.25-(-2.375)]}{2}=-2.375+0.0625=\ -2.3125$$

$$f(x_4)=f(-2.3125)=2(-2.3125)^2+5(-2.3125)+1=+0.1328$$

Interval Analysis:

$$\left[\frac{-2.375}{f(a)=+} \quad \frac{-2.3125}{f(x_4)=+} \quad \frac{-2.25}{f(b)=-}\right]$$

$f(a)*f(x_4)>0$, in the Interval [-2.375, -2.3125], does not cut the graph to the X-axis, therefore, it is not the object of study.

$f(x_4)*f(b)<0$, in the Interval [-2.3125, -2.25], if it cuts the graph to the X-axis, therefore, if it is the object of study.

Percentage error up to Iteration 4.

$$E_p=\left|\frac{V_{act}-V_{ant}}{Vact}\right|*100$$

$$E_p=\left|\frac{-2.3125-(-2.375)}{-2.3125}\right|*100=2.70\%$$

Conclusions of the method:

In iteration No. 4 the function f(x) crosses the x-axis at approximately x = -2.3125, however, according to the percentage error there is a 2.70% error with respect to the true value, i.e. x = -2.3125 is not the true value, but an approximate value.

The value closest to 0% is obtained in iteration number 12; for practical reasons the remaining seven iterations are not developed, but it is made clear that they are carried out in the same way as the first four.

Iteration 12.

In iteration No. 11 it was concluded with an interval [-2.2813, -2.28], where
a = -2.2813 - b = -2.28 y $x_{11} = -2.2808$

Consider the expression again: $x_i = a + (b - a)/2$

$$x_{12} = -2.2813 + \frac{[-2.28 - (-2.2813)]}{2} = -2.2813 + 0.0007$$
$$= -2.2806$$

Percentage error up to Iteration 12.

$$E_p = \left|\frac{-2.2806 - (-2.2808)}{-2.2806}\right| * 100 = 0.0088\%$$

It is not necessary to evaluate f(x_{12}) as the Percentage Error (Ep) is close to 0%, therefore, it follows that the value of f(x) when it crosses over the X-axis is -2.2806.

Table 1 is created in an Excel sheet, and contains all the values obtained from each iteration until reaching 0%. (a is the left end of the interval, b is the right end of the interval, Vm = Xi is the mean value of the intervals; f(a), f(xi) and f(b), are the variable evaluations, Ea is absolute error, Er is relative error and Ep is the percentage error).

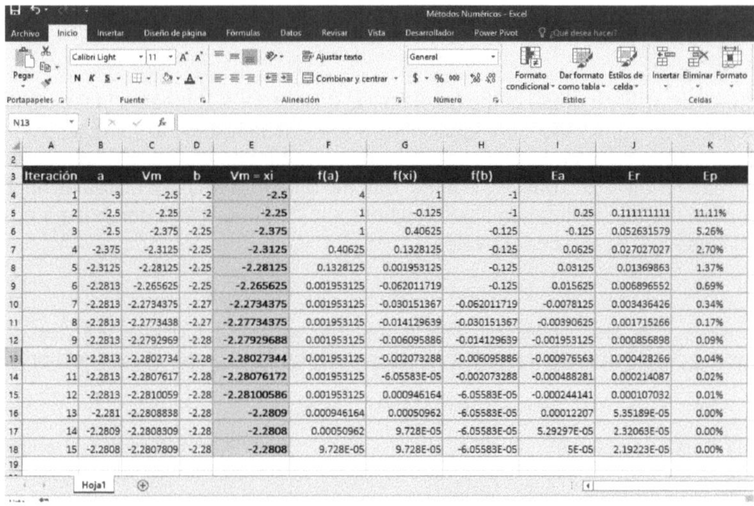

Iteración	a	Vm	b	Vm = xi	f(a)	f(xi)	f(b)	Ea	Er	Ep
1	-3	-2.5	-2	**-2.5**	4	1	-1			
2	-2.5	-2.25	-2	**-2.25**	1	-0.125	-1	0.25	0.111111111	11.11%
3	-2.5	-2.375	-2.25	**-2.375**	1	0.40625	-0.125	-0.125	0.052631579	5.26%
4	-2.375	-2.3125	-2.25	**-2.3125**	0.40625	0.1328125	-0.125	0.0625	0.027027027	2.70%
5	-2.3125	-2.28125	-2.25	**-2.28125**	0.1328125	0.001953125	-0.125	0.03125	0.01369863	1.37%
6	-2.2813	-2.265625	-2.25	**-2.265625**	0.001953125	-0.062011719	-0.125	0.015625	0.006896552	0.69%
7	-2.2813	-2.2734375	-2.27	**-2.2734375**	0.001953125	-0.030151367	-0.062011719	-0.0078125	0.003436426	0.34%
8	-2.2813	-2.2773438	-2.27	**-2.27734375**	0.001953125	-0.014129639	-0.030151367	-0.00390625	0.001715266	0.17%
9	-2.2813	-2.2792969	-2.28	**-2.27929688**	0.001953125	-0.006095886	-0.014129639	-0.001953125	0.000856898	0.09%
10	-2.2813	-2.2802734	-2.28	**-2.28027344**	0.001953125	-0.002073288	-0.006095886	-0.000976563	0.000428266	0.04%
11	-2.2813	-2.2807617	-2.28	**-2.28076172**	0.001953125	-6.05583E-05	-0.002073288	-0.000488281	0.000214087	0.02%
12	-2.2813	-2.2810059	-2.28	**-2.28100586**	0.001953125	0.000946164	-6.05583E-05	-0.000244141	0.000107032	0.01%
13	-2.281	-2.2808838	-2.28	**-2.2809**	0.000946164	0.00050962	-6.05583E-05	0.00012207	5.35189E-05	0.00%
14	-2.2809	-2.2808309	-2.28	**-2.2808**	0.00050962	9.728E-05	-6.05583E-05	5.29297E-05	2.32063E-05	0.00%
15	-2.2808	-2.2807809	-2.28	**-2.2808**	9.728E-05	9.728E-05	-6.05583E-05	5E-05	2.19223E-05	0.00%

Table 1. Iterations of the function f(x)

1.2 GRAPHICAL METHOD.

Figure 1 shows the graph of the function $f(x) = 2x^2 + 5x + 1$, created in the classic Geogebra 5.0 application; the application shows visually that the value where the function f(x) crosses over the x-axis is indeed in the interval [-3,-2], with x being no greater than -2.5. Thus, we confirm that the data that was calculated using the bisection method in iteration No. 12, where x = - 2.2806 with a percentage error (Ep) close to 0%, is a correct value.

Likewise, it is also observed that in the interval [-1,0] f(x) crosses over the abscissa, while in the interval [1,2] it does not cross the graph of the function.

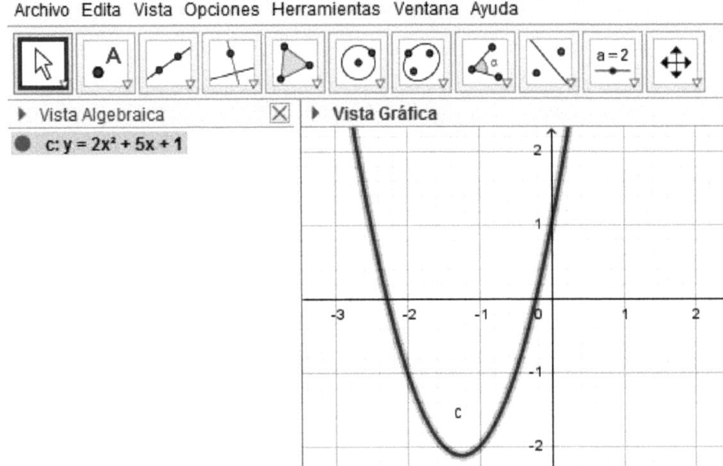

Figure No. 1 Graph of function 2x +5x+1[2]

1.3 FALSE RULE METHOD

Like the method above, this, too, is considered a closed interval method, requiring two initial values for its root; although the bisection method is a perfectly valid technique for determining roots, its "brute force" approximation method is relatively inefficient. The false position is an alternative based on finite iterations.

This method determines roots of x that satisfy an equation f(x), where, in most cases f(x) will be a known, real, one-variable, continuous function, in fact, once or twice differentiable in the region of interest. The function f(x) may be a polynomial.

In this topic we give a real function f of a real variable x, examine its behaviour for a very large magnitude of x and thus locate all its roots within one of its finite intervals.

To demonstrate and achieve a comparison between methods, the same function performed in the bisection method is developed with the same intervals, but the reader is left to construct their own criteria.

Given the function f(x) = $2x^2 + 5x + 1$

a) Determine in which of the following intervals [1,2], [-3,-2] and [-1,0] is the value of the variable **X** such that the graph of the function f(x) crosses the X-axis.
b) What is the value of X at the point where the graph crosses the X-axis.

Procedure to determine in which interval the graph crosses.

The procedure for determining the study intervals are exactly the same as those used in the **Bisection Method**. They are described below:

• ***For the interval [1,2], considering the interval [a, b], where a = 1 and b = 2***

• Evaluate f(a) and f(b):

$$f(a) = f(1) = 2(1)^2 + 5(1) + 1 = 8$$
$$f(b) = f(2) = 2(2)^2 + 5(2) + 1 = 19$$

• If f(a)*f(b) < 0, then, the graph passes in the studied interval through the X-axis, otherwise it does not cut the axis. In this case:

•

$$f(a) * f(b) = 152, es\ mayor\ que\ CERO.$$

Therefore, in the interval [1,2] it does not cross the graph of the function on the X axis and is not the object of study as it does not form part of the function.

• ***For the interval [-3,-2], considering the interval [a, b], where a = -3 and b = -2***

• Evaluate f(a) and f(b):

$$f(a) = f(-3) = 2(-3)^2 + 5(-3) + 1 = 4$$
$$f(b) = f(-2) = 2(-2)^2 + 5(-2) + 1 = -1$$

• Since f(a)*f(b) < 0, then, the graph passes through the studied interval crossing the X-axis. In this case:

$$f(a) * f(b) = -4, es\ menor\ que\ CERO.$$

Therefore, in the interval [-3,-2] it does cross the graph of the function on the X axis and is the object of study as it forms part of the function.

• ***For the interval [-1,0], considering the interval [a, b], where a = -1 and b = 0***
• Evaluate f(a) and f(b):

$$f(a) = f(-1) = 2(-1)^2 + 5(-1) + 1 = -2$$
$$f(b) = f(0) = 2(0)^2 + 5(0) + 1 = +1$$

• Since f(a)*f(b) < 0, then, the graph passes through the studied interval crossing the X-axis. In this case:

$$f(a) * f(b) = -2, es\ menor\ que\ CERO.$$

Therefore, in the interval [-1,0] it does cross the graph of the function on the X axis and is the object of study as it forms part of the function.

Analysis process of each interval by iterations seeking to calculate the value of X as close as possible, where X = Xi, when i = 1, 2, 3 ... n

First interval to be studied [-3, -2].
Determine the value that exists between [-3, -2] using the following formula:

$x_i = \frac{af(b)-bf(a)}{f(b)-f(a)}$(Formula No. 3) then:

Iteration 1.
Where: a = -3 and b = -2, evaluating these values in the function, we have:

$$f(a) = f(-3) = 2(-3)^2 + 5(-3) + 1 = +4$$
$$f(b) = f(-2) = 2(-2)^2 + 5(-2) + 1 = -1$$

With these values evaluated in the function and those of the interval, they are substituted in the formula No. 3

$$x_1 = \frac{(-3)(-1) - (-2)(4)}{-1 - (+4)} = \frac{3 + 8}{-5} = \frac{11}{-5} = -2.2$$

Evaluating f(xi):

$$f(x_1) = f(-2.2) = 2(-2.2)^2 + 5(-2.2) + 1 = -0.32$$

Interval Analysis:

$$\left[\frac{-3}{f(a)=+} \quad \frac{-2.2}{f(x_1)=-} \quad \frac{-2}{f(b)=-}\right]$$

As $f(a) * f(x_1) < 0$, then, in the Interval [-3, -2.2], it cuts the graph to the X-axis, therefore, it is the object of study; reciprocally:

$f(x_1) * f(b) > 0$, in the Interval [-2.2, -2], then, it does not cut the graph to the X-axis, therefore, it is not the object of study.

Iteration 2.
In the interval [-3, -2.2], where a = -3 and b = -2.2
Evaluating these values in the function, one has:

$$f(a) = f(-3) = 2(-3)^2 + 5(-3) + 1 = +4$$
$$f(b) = f(-2.2) = 2(-2.2)^2 + 5(-2.2) + 1 = -0.32$$

Considering formula No. 1

$$x_2 = \frac{(-3)(-0.32) - (-2.2)(+4)}{-0.32 - 4} = \frac{0.96 + 8.8}{-4.32} = -2.2593$$

$$f(x_2) = f(-2.2593) = 2(-2.2593)^2 + 5(-2.2593) + 1 = -0.0876$$

Interval Analysis:

$$\left[\frac{-3}{f(a)=+} \quad \frac{-2.2593}{f(x_2)=-} \quad \frac{-2.2}{f(b)=-}\right]$$

$f(a) * f(x_2) < 0$, in the Interval [-3, -2.2593], if it cuts the graph to the X-axis, therefore, if it is the object of study.

$f(x_2) * f(b) > 0$, in the Interval [-2.2593, -2.2], does not cut the graph to the X-axis, therefore, it is not the object of study.

Percentage error up to Iteration 2.

$$E_p = \left|\frac{V_{act} - V_{ant}}{Vact}\right| * 100 \quad Fórmula\ No.2$$

Where:

V_{act} en Current value
V_{ant} en Previous value
Ep en Percentage error (%)

$$E_p = \left|\frac{-2.2593 - (-2.2)}{-2.2593}\right| * 100 = 2.6247\%$$

Iteration 3.

In the interval [-3, -2.2593], where a = -3 and b = -2.2593
Evaluating these values in the function, one has:

$f(a) = f(-3) = 2(-3)^2 + 5(-3) + 1 = +4$
$f(b) = f(-2.2593) = 2(-2.2593)^2 + 5(-2.2593) + 1 = -0.0876$

Considering formula No. 1

$$x_3 = \frac{(-3)(-0.0876) - (-2.2593)(+4)}{-0.0876 - 4} = \frac{9.2952}{-4.0876} = -2.2740$$

$$f(x_3) = f(-2.2740) = 2(-2.2740)^2 + 5(-2.2740) + 1 = -0.027848$$

Interval Analysis:

$$\left[\frac{-3}{f(a) = +} \quad \frac{-2.2740}{f(x_3) = -} \quad \frac{-2.2593}{f(b) = -}\right]$$

$f(a) * f(x_3) < 0$, in the Interval [-3, -2.2740], if it cuts the graph to the X-axis, therefore, it is the object of study .[1]

$f(x_3) * f(b) > 0$, in the Interval [-2.2740, -2.2593], does not cut the graph to the X-axis, therefore, it is not the object of study.

Percentage error up to Iteration 3.

$$E_p = \left|\frac{-2.2740 - (-2.2593)}{-2.2740}\right| * 100 = 0.65\%$$

With the percentage error of 0.65%, the root value at **x = -2.2740** can already be considered, however, if you want to get as close to ZERO percent as possible, it is necessary to continue the iterations by making the procedure repetitive similar to the previous iterations.

[1] Value used for Iteration No. 4, the value in the excel sheet is not considered.

See excel table 2 below.

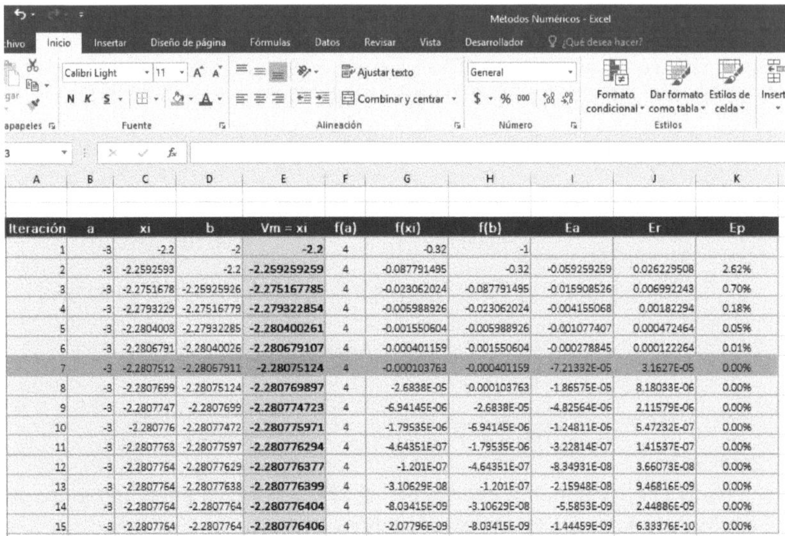

Iteración	a	xi	b	Vm = xi	f(a)	f(xi)	f(b)	Ea	Er	Ep
1	-3	-2.2	-2	**-2.2**	4	-0.32	-1			
2	-3	-2.2592593	-2.2	**-2.259259259**	4	-0.087791495	-0.32	-0.059259259	0.026229508	2.62%
3	-3	-2.2751678	-2.25925926	**-2.275167785**	4	-0.023062024	-0.087791495	-0.015908526	0.006992243	0.70%
4	-3	-2.2793229	-2.27516779	**-2.279322854**	4	-0.005988926	-0.023062024	-0.004155068	0.00182294	0.18%
5	-3	-2.2804003	-2.27932285	**-2.280400261**	4	-0.001550604	-0.005988926	-0.001077407	0.000472464	0.05%
6	-3	-2.2806791	-2.28040026	**-2.280679107**	4	-0.000401159	-0.001550604	-0.000278845	0.000122264	0.01%
7	-3	-2.2807512	-2.28067911	**-2.28075124**	4	-0.000103763	-0.000401159	-7.21332E-05	3.1627E-05	0.00%
8	-3	-2.2807699	-2.28075124	**-2.280769897**	4	-2.6838E-05	-0.000103763	-1.86575E-05	8.18033E-06	0.00%
9	-3	-2.2807747	-2.2807699	**-2.280774723**	4	-6.94145E-06	-2.6838E-05	-4.82564E-06	2.11579E-06	0.00%
10	-3	-2.280776	-2.28077472	**-2.280775971**	4	-1.79535E-06	-6.94145E-06	-1.24811E-06	5.47232E-07	0.00%
11	-3	-2.2807763	-2.28077597	**-2.280776294**	4	-4.64351E-07	-1.79535E-06	-3.22814E-07	1.41537E-07	0.00%
12	-3	-2.2807764	-2.28077629	**-2.280776377**	4	-1.201E-07	-4.64351E-07	-8.34931E-08	3.66073E-08	0.00%
13	-3	-2.2807764	-2.28077638	**-2.280776399**	4	-3.10629E-08	-1.201E-07	-2.15948E-08	9.46816E-09	0.00%
14	-3	-2.2807764	-2.2807764	**-2.280776404**	4	-8.03415E-09	-3.10629E-08	-5.5853E-09	2.44886E-09	0.00%
15	-3	-2.2807764	-2.2807764	**-2.280776406**	4	-2.07796E-09	-8.03415E-09	-1.44459E-09	6.33376E-10	0.00%

Table 2 Excel sheet-False Position Method

Note that in the:

Iteration 4.

With an interval [-3, -2.2740], where a = -3 and b = -2.2740, we obtain a:

x_4 =-2.2793 and Ep = 18%.

Iteration 5.

With an interval [-3, -2.2793], where a = -3 and b = -2.2793, we obtain a:

x_5 =-2.2804 and Ep = 0.05%.

Iteration 6.

With an interval [-3, -2.2804], where a = -3 and b = -2.2804, one obtains an:

$x_6 = -2.2807$[2] and an Ep = 0.01%.

While in the:

Iteration 7.

With an interval [-3, -2.2807], where a = -3 and b = -2.2807, one obtains an:

x_7 =-2.2807 and an Ep = 0.00%.

Conclusions of the method:

It is concluded that the value of the root of the function that is closest to 0% is given at Iteration No. 7. This value of x_7 can also be checked through the Bisection Method and the Graphical Method seen previously.

So, X_7 = -2.2807

In the same way, the interval [-1, 0] can be evaluated to obtain precisely the second point where the function f(x) intersects the X-axis.

[2] Value taken from Figure 1, but rounded to 10 thousandth of a position.

Chapter 2

Closed interval methods

2.1 Fixed point method.
2.2 Newton Raphson method
2.3 Secant method
2.4 Multiple Root Method.

2.1 FIXED POINT METHOD.

The Fixed Point Method is a numerical resolution technique that is classified as an open interval method. This approach uses systematic trial-and-error iterations and is particularly useful when closed methods do not provide satisfactory results. Although computationally efficient, it does not always guarantee convergence to a solution (Chapra & P. Canale, 2007).

This method is used to approximate the root of a non-linear equation. Its formulation is based on the restructuring of an equation of the form $f(x) = 0$. Through this iteration, one seeks to find a value of x that remains on the left-hand side of the equation.

In this anthology, the steps necessary to carry out a correct iteration are explained in detail, thus guaranteeing a satisfactory result. The same function discussed in the Bisection and False Position Methods will be used, with the aim of demonstrating that it is possible to achieve the same result through the Fixed Point Method. I hope you find this information useful.

Unlike the two previous methods, with the fixed point method we can only find one of the two roots of the function f(x), hence its name, since the interval studied is so wide that it goes $(-\infty, +\infty)$ and is not able to find more than one root.

Recall that the function f(x) = $2x^2 + 5x + 1$ has two roots and they are in the intervals [-3,-2] and [-1,0]; so let's see what happens when we develop this method.

Procedure for determining the root value.

- First. We equal the function to zero:

$$f(x) = 2x^2 + 5x + 1$$

$$2x^2 + 5x + 1 = 0$$

- Then, we identify the variable with the lowest degree in the function and we clear it.

$$2x^2 + 5x + 1 = 0$$

$$5x = -2x^2 - 1$$

$$x = \frac{-2x^2 - 1}{5}$$

- Once the clearance or equivalence has been determined, the next step is to test the root[3] . It is recommended that the initial value of the first iteration x_0 is equal to 0 (zero). Thus: $x_i = x_0 = 0$, and is the value that the quadratic variable assumes in the clearance, hence:

- The following iterations are calculated as follows:

$$x_{i+1} = \frac{-2x^2 - 1}{5}$$

$$x_1 = \frac{-2(0)^2 - 1}{5} = -\frac{1}{5} = -0.2$$

$$x_2 = \frac{-2(-0.2)^2 - 1}{5} = -\frac{27}{50} = -0.216$$

[3] Numerical Methods Applied to Engineering

Considering that the Percentage Error (Ep) is determined by the formula 2

$$E_p = \left|\frac{-0.216 - (-0.2)}{-0.216}\right| * 100 = 7.4074\%$$

$$x_3 = \frac{-2(-0.216)^2 - 1}{5} = -0.2187$$

$$E_p = \left|\frac{-0.2187 - (-0.216)}{-0.2187}\right| * 100 = 1.2346\%$$

In iteration No. 4, it can be observed that the value of the root can already be considered as a true value, since it is the one that is closest to 0% of Ep, therefore, it can already be considered as the value that intersects the function f(x) with the **X-axis**. Observe.

$$x_4 = \frac{-2(-0.2187)^2 - 1}{5} = -0.2191$$

$$E_p = \left|\frac{-0.2191 - (-0.2187)}{-0.2191}\right| * 100 = 0.1971\%$$

Whereas in iteration 5:

$$x_5 = \frac{-2(-0.2191)^2 - 1}{5} =$$

$$E_p = \left|\frac{-0.2192 - (-0.2191)}{-0.2192}\right| * 100 = 0.0456\%$$

With the value obtained in iteration No. 5, where x_5 = -0.2192 it is observed that a percentage error too close to 0% is achieved, under this argument, it can be stated that the root of the function is in the interval [-1, 0] and decides to conclude the iterations.
The function is then checked to demonstrate the veracity of the x-$value_5$ = -0.2192.

- **Check.**

Substitution of the x-value$_5$ = -0.2192 in the function:

$$f(x) = 2x^2 + 5x + 1$$

$$2(-0.2192)^2 + 5(-0.2192) + 1 = 0$$

$$0.0961 - 1.096 + 1 = 0$$

$$0.0001 \cong 0$$

The following Excel table 3 shows all the possible iterations for this function, as well as the variation and approximation of the Ep.

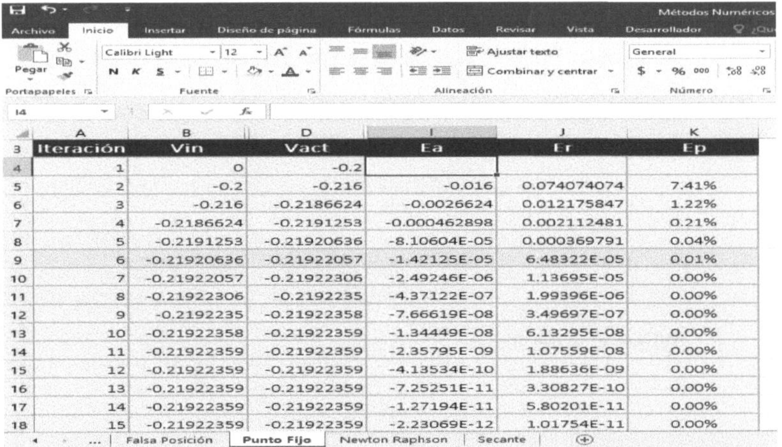

Iteración	Vin	Vact	Ea	Er	Ep
1	0	-0.2			
2	-0.2	-0.216	-0.016	0.074074074	7.41%
3	-0.216	-0.2186624	-0.0026624	0.012175847	1.22%
4	-0.2186624	-0.2191253	-0.000462898	0.002112481	0.21%
5	-0.2191253	-0.21920636	-8.10604E-05	0.000369791	0.04%
6	-0.21920636	-0.21922057	-1.42125E-05	6.48322E-05	0.01%
7	-0.21922057	-0.21922306	-2.49246E-06	1.13695E-05	0.00%
8	-0.21922306	-0.2192235	-4.37122E-07	1.99396E-06	0.00%
9	-0.2192235	-0.21922358	-7.66619E-08	3.49697E-07	0.00%
10	-0.21922358	-0.21922359	-1.34449E-08	6.13295E-08	0.00%
11	-0.21922359	-0.21922359	-2.35795E-09	1.07559E-08	0.00%
12	-0.21922359	-0.21922359	-4.13534E-10	1.88636E-09	0.00%
13	-0.21922359	-0.21922359	-7.25251E-11	3.30827E-10	0.00%
14	-0.21922359	-0.21922359	-1.27194E-11	5.80201E-11	0.00%
15	-0.21922359	-0.21922359	-2.23069E-12	1.01754E-11	0.00%

Table 3 Excel worksheet-Fixed Point Method

Conclusions of the method:

The method is easy, although it does not always work and does not always give a completely effective answer, as is the case for the second value in the interval [-3, -2].

You only need basic knowledge of clearings, substitutions and of course the use of a calculator or at best the powerful excel.

2.2 NEWTON RAPHSON METHOD.

The Newton-Raphson method is a widely used numerical technique for finding the roots of nonlinear equations (Chapra & P. Canale, 2007). Classified as an open interval method, this method is characterised by its speed and efficiency in converging to a solution. It often outperforms other methods, such as the bisection method and those described above.

To successfully apply this method, it is essential to have a basic knowledge of algebra, calculus of derivatives, evaluation of functions and handling of percentages. In this anthology, we detail the steps necessary to perform correct iterations, ensuring satisfactory results. We will use the same function approached in the previous methods, with the aim of demonstrating that the same result can be achieved. I hope you find this guide useful.

Perhaps of all the formulas for calculating roots, Newton Raphson's expression is the most widely used. If the initial value for the root is x_i , then a tangent can be drawn from the point [x_i , f(x_i)] on the curve of a function, so the point where this tangent line intersects the x-axis represents an improved approximation of the root .[4]

With this method you can determine more than one differential root than with the fixed point method.

Recall that the function $f(x) = 2x^2 + 5x + 1$ has two roots and they are in the intervals [-3,-2] and [-1,0]; so let's see what happens when we develop this method.

Procedure for determining the root value.

- First. The Newton Raphson formula is proposed:

$$x_{i+1} = x_i - \frac{f(x_i)}{f'(x_i)} \qquad Fórmula\ No. 4$$

  Note that to use the formula you need an initial value, evaluate and derive a function.
- One of the questions that sometimes needs to be answered in mathematics is, what values do we need to give to x in a function f(x) for it to be 0, i.e. f(x) = 0? If we manage to obtain an exact or very approximate answer, whatever the means by which we have obtained it, we will then have the points of a function that will intersect the x-axis in a Cartesian plane. Let us begin by deriving the function f(x):

$$f(x) = 2x^2 + 5x + 1$$
$$f'(x) = 4x + 5$$

- Ready, now we can start the iterations; the first one, known as x_0 can be any value, and assigned in an arbitrary way, I recommend starting with $x_i = x_0 = 0$.

  Iteration No. 1. Using Newton Raphson's formula we substitute the initial value of $x_0 = 0$

$$x_{i+1} = x_i - \frac{f(x_i)}{f'(x_i)}$$

[4] Numerical Methods for Engineers. Steven C. Chapra. Fifth Edition. Mc Graw Hill

Evaluating f(x) and f'(x), we have:

$$f(0) = 2(0)^2 + 5(0) + 1$$
$$f(0) = 1$$
$$f'(0) = 4(0) + 5$$
$$f'(0) = 5$$

With the values obtained, they are substituted into formula No. 4:

$$x_1 = (0) - \frac{1}{5}$$

$$x_1 = -\frac{1}{5} = -0.20$$

Iteration No. 2. With the value of x_1 = -0.20, the second iteration is determined; evaluating f(x) and f'(x), we have:

$$f(-0.20) = 2(-0.20)^2 + 5(-0.20) + 1$$

$$f(-0.20) = 0.08 - 1 + 1$$

$$f(0.20) = 0.08$$

$$f'(-0.20) = 4(-0.20) + 5$$

$$f'(-0.20) = -0.80 + 5$$

$$f'(-0.20) = 4.20 \therefore$$

With the values obtained, they are substituted into formula No. 4:

$$x_2 = -0.20 - \frac{0.08}{4.20}$$

$$x_2 = -0.20 - 0.0190 = -0.2190$$

Considering that the Percentage Error (Ep) is determined with the following formula, it is obtained:

$$E_p = \left|\frac{V_{act} - V_{ant}}{Vact}\right| * 100 \quad Fórmula\ No. 2$$

Where: V_{act} is Present value
V_{ant} en Previous value
Ep en Percentage error (%)

$$E_p = \left|\frac{-0.2190 - (-0.20)}{-0.2190}\right| * 100 = 8.6758\%$$

Iteration No. 3. With the value of x_2 = -0.2190, the third iteration is determined by evaluating f(x) and f'(x) and arrives at:

$$f(-0.2190) = 2(-0.2190)^2 + 5(-0.2190) + 1$$

$$f(-0.2190) = 0.09592 - 1.095 + 1$$

$$f(-0.2190) = 0.00092$$

$$f'(-0.2190) = 4(-0.2190) + 5$$

$$f'(-0.2190) = -0.8760 + 5$$

$$f'(-0.2190) = 4.1240 \therefore$$

With the values obtained, they are substituted into formula No. 4:

$$x_3 = -0.2190 - \frac{0.00092}{4.1240}$$

$$x_3 = -0.2190 - 0.0002231 = -0.2192$$

Determining the Percentage Error (Ep) with the formula No. 2, it is obtained:

$$E_p = \left|\frac{-0.2192 - (-0.2190)}{-0.2192}\right| * 100 = 0.0912\%$$

With the value obtained in iteration No. 3, where x_3 = -0.2192 it is observed that the percentage error is too close to 0%, under this argument, it can be stated that the root of the function is in the interval [-1, 0] and it is decided to conclude the iterations.
The function is then checked to demonstrate the veracity of the value x_3 = -0.2192.

- **Check.**
  Substitution of the x-value$_3$ = -0.2192 in the function:

  f(x) = $2x^2 + 5x + 1$

  $2(-0.2192)^2 + 5(-0.2192) + 1 = 0$

  $0.0961 - 1.096 + 1 = 0$

  $0.0001 \cong 0$

The following excel table 4 shows all the possible iterations for this function, as well as the variation and approximation obtained on the Ep.

Iteración	Xn	Xn+1	f(x)	f'(x)	Ea	Er	Ep
1	0	-0.2	1	5			
2	-0.2	-0.219047619	0.08	4.2	-0.2	1	100.00%
3	-0.2190476	-0.219223579	0.000725624	4.123809524	-0.019047619	0.086956522	8.70%
4	-0.2192236	-0.219223594	6.19235E-08	4.123105686	-0.00017596	0.000802649	0.08%
5	-0.2192236	-0.219223594	0	4.123105626	-1.50187E-08	6.85084E-08	0.00%
6	-0.2192236	-0.219223594	0	4.123105626	0	0	0.00%
7	-0.2192236	-0.219223594	0	4.123105626	0	0.000000E+00	0.00%
8	-0.2192236	-0.219223594	0	4.123105626	0	0.000000E+00	0.00%
9	-0.2192236	-0.219223594	0	4.123105626	0	0.000000E+00	0.00%
10	-0.2192236	-0.219223594	0	4.123105626	0	0.000000E+00	0.00%
11	-0.2192236	-0.219223594	0	4.123105626	0	0.000000E+00	0.00%
12	-0.2192236	-0.219223594	0	4.123105626	0	0.000000E+00	0.00%
13	-0.2192236	-0.219223594	0	4.123105626	0	0.000000E+00	0.00%
14	-0.2192236	-0.219223594	0	4.123105626	0	0.000000E+00	0.00%
15	-0.2192236	-0.219223594	0	4.123105626	0	0.000000E+00	0.00%

Falsa Posición | Punto Fijo | Newton Raphson | Secante

Table 4. Excel sheet-Newton Raphson's method

The difference between this method and the fixed point method, as mentioned at the beginning, is that you can calculate both roots, since the initial value of x_0 does not necessarily have to be ZERO; the value of the root that lies in the interval [-3, -2] can be found if you assign x_0 the value of -3; I'll leave it as homework!

If you formulate in an Excel sheet the development of this method you will be able to check both values of x_0 that I have recommended to you.

CONCLUSIONS OF THE METHOD:

The method is easy and less complex than the fixed point method, and is valid for calculating more than one root, which makes it different from other open interval methods. As can be seen, it is important to know how to derive, otherwise this could be an obstacle to reach the correct result. I recommend running the procedures in an Excel spreadsheet to know the different values of x_i that are obtained for each iteration...

2.3 SECANT METHOD.

The Secant Method is a powerful and efficient technique for finding the zeros of a function iteratively. Unlike closed methods which require the function to change sign between two points, this is an Open Interval Method which relies on two initial values of x. This feature allows the method to be applied in situations where the sign change condition is not met, which extends its usefulness in practice.

In contrast to the Newton-Raphson Method, which uses the derivative of the function to calculate the approximation to the root, the Secant Method relies on the extrapolation of a secant through two points in the function f(x). By employing a divided difference instead of a derivative, this method simplifies the calculation process, making it accessible even in cases where the derivative may be complicated or costly to obtain (Nieves & C. Domínguez, 1998).

The following paragraphs and calculations aim to guide you through the steps necessary to perform a successful iteration with the Secant Method, using the same function presented in previous methods. In doing so, we seek to demonstrate that it is possible to achieve consistent and satisfactory results, regardless of the approach used. We hope you find this information useful in your learning and application of numerical methods.

Among the formulas for calculating roots, the Newton-Raphson method is probably the most widely used. However, its main drawback is that it requires knowing the value of the first derivative of the function at the specific point. In some cases, the shape of f(x) makes it difficult to calculate this derivative. In such situations, the Secant Method presents itself as a more favourable alternative, since it is based on two values evaluated directly on the function without the need to modify it. From these values, the root is determined using a divided difference.

The function f(x) to be developed for this method is the same as the one we have worked on in the previous topics: $f(x) = 2x^2 + 5x + 1$[-3,-2], so we already know that it has two roots and they are in the intervals [-3,-2] and [-1,0]; so let's see what happens when we develop this method.

Procedure for determining the root value.

- First. The secant formula is posed:

$$x_i = x_{i-1} - \frac{f(x_{i-1})(x_{i-1} - x_{i-2})}{f(x_{i-1}) - f(x_{i-2})} \qquad Fórmula\ No.5$$

- Let us consider the function f(x) under study:

$$f(x) = 2x^2 + 5x + 1$$

- Ready, let's start the iterations; so it is necessary to define the two initial values; it can be any, I recommend that this time we start with:

$$x_0 = -2 \qquad x_1 = 1$$

These values are chosen with reference to the two intervals we already know, but they can ultimately be any.

Iteration No. 2. Using the secant formula we substitute the initial values. Knowing that: $x_0 = -2 \qquad x_1 = 1$

Evaluating f(x) we have:

$$f(x_0) = f(-2) = 2(-2)^2 + 5(-2) + 1$$
$$f(-2) = 8 - 10 + 1$$
$$f(-2) = -1$$
$$f(x_1) = f(1) = 2(1)^2 + 5(1) + 1$$
$$f(x_i) = 2 + 5 + 1$$
$$f(0) = 8$$

With the values obtained, they are substituted into formula No. 5:

$$x_i = x_{i-1} - \frac{f(x_{i-1})(x_{i-1} - x_{i-2})}{f(x_{i-1}) - f(x_{i-2})}$$

$$x_2 = x_1 - \frac{f(x_1)(x_1 - x_0)}{f(x_1) - f(x_0)}$$
$$x_2 = 1 - \frac{8[1 - (-2)]}{8 - (-1)}$$

$$x_2 = 1 - \frac{24}{9} = -\frac{5}{3} = -1.6666, \textit{este valor es el nuevo } x_i$$

Iteration No. 3. With the value of x_2 = -1.6666, the third iteration is determined considering that: $x_1 = 1,\ x_2 = -1.6666$

$$x_3 = x_2 - \frac{f(x_2)(x_2 - x_1)}{f(x_2) - f(x_1)}$$

Evaluating f(x) we have:

$$f(x_1) = f(1) = 2(1)^2 + 5(1) + 1$$
$$f(1) = 2 + 5 + 1$$
$$f(1) = 8$$

$$f(x_2) = f(-1.6666) = 2(-1.6666)^2 + 5(-1.6666) + 1$$
$$f(-1.6666) = 5.5551 - 8.333 + 1$$
$$f(-1.6666) = -1.7777$$

Using the secant formula and substituting:

$$x_3 = -\frac{5}{3} - \frac{-1.7777\left(-\frac{5}{3} - 1\right)}{-1.7777 - 8}$$

$$x_3 = -\frac{5}{3} - \frac{\frac{128}{27}}{-\frac{88}{9}} = -\frac{5}{3} - \left(-\frac{16}{33}\right)$$
$$x_3 = -\frac{13}{11} = -1.1818, \textit{este valor es el nuevo } x_i$$

Considering that the Percentage Error (Ep) is determined with the following formula, it is obtained:

$$E_p = \left|\frac{V_{act} - V_{ant}}{Vact}\right| * 100 \quad \textit{Fórmula No.2}$$

Where: V_{act} is Present value
V_{ant} en Previous value
Ep en Percentage error (%)

$$E_p = \left|\frac{-1.1818 - (-1.6666)}{-1.1818}\right| * 100 = 41.02\%$$

The root value is still far removed from the true value, that is what the percentage error indicates. E_p.

Iteration No. 4. With the value of x_3 = -1.1818, the next iteration is determined considering that: $x_2 = -1.6666,\ x_3 = -1.1818$

$$x_4 = x_3 - \frac{f(x_3)(x_3 - x_2)}{f(x_3) - f(x_2)}$$

Evaluating f(x) we have:

$$f(x_2) = f(-1.6666) = 2(-1.6666)^2 + 5(-1.6666) + 1$$
$$f(-1.6666) = 5.5551 - 8.333 + 1$$
$$f(-1.6666) = -1.7779$$

$$f(x_3) = f(-1.1818) = 2(-1.1818)^2 + 5(-1.1818) + 1$$
$$f(-1.1818) = 2.7933 - 5.909 + 1$$
$$f(-1.1818) = -2.1157$$

Using the secant formula and substituting:

$$x_4 = -1.1818 - \frac{(-2.1157)[-1.1818 - (-1.6666)]}{-2.1157 - (-1.7779)}$$

$$x_4 = -1.1818 - \frac{768}{253} = -4.2174, este\ valor\ es\ el\ nuevo\ x_i$$

Determining the Percentage Error (Ep) with the formula No. 2, it is obtained:

$$E_p = \left|\frac{-4.2174 - (-1.1818)}{-4.2174}\right| * 100 = 71.98\%$$

With the value obtained in iteration No. 4, where x_4 = -4.2174, we observe a result in the percentage error that could ***scare us*** and turn on the yellow warning lights, because far from decreasing the percentage (that is, getting closer to the real value of the root), it has increased; I recommend keeping calm and giving the necessary confidence to the method.

Iteration No. 5. With the value of x_4 = -4.2174, the next iteration is determined considering that: $x_3 = -1.1818,\ x_4 = -4.2174$

$$x_5 = x_4 - \frac{f(x_4)(x_4 - x_3)}{f(x_4) - f(x_3)}$$

Evaluating f(x) we have:

$$f(x_3) = f(-1.1818) = 2(-1.1818)^2 + 5(-1.1818) + 1$$
$$f(-1.1818) = 2.7933 - 5.909 + 1$$
$$f(-1.1818) = -2.1157$$

$$f(x_4) = f(-4.2174) = 2(-4.2174)^2 + 5(-4.2174) + 1$$
$$f(-4.2174) = 35.5729 - 21.087 + 1$$
$$f(-4.2174) = 15.4859$$

Using the secant formula and substituting:

$$x_5 = -4.2174 - \frac{(15.4859)[-4.2174 - (-1.1818)]}{15.4859 - (-2.1157)}$$

$$x_5 = -4.2174 - \frac{(-47.0090)}{17.6016} = -1.5467, este\ valor\ es\ el\ nuevo\ x_i$$

Determining the Percentage Error (Ep) with the formula No. 2, it is obtained:

$$E_p = \left|\frac{-1.5467 - (-4.2174)}{-1.5467}\right| * 100 = 172.67\%$$

I recommend not to panic, even if the percentage error continues to increase; at some point in the iterations it will become stable, much depends on the values chosen initially.

Iteration No. 6. With the value of x_4 = -4.2174, the next iteration is determined considering that: $x_4 = -4.2174,\ x_5 = -1.5467$

$$x_6 = x_5 - \frac{f(x_5)(x_5 - x_4)}{f(x_5) - f(x_4)}$$

Evaluating f(x) we have:

$$f(x_4) = f(-4.2174) = 2(-4.2174)^2 + 5(-4.2174) + 1$$
$$f(-4.2174) = 35.5729 - 21.087 + 1$$
$$f(-4.2174) = 15.4859$$

$$f(x_5) = f(-1.5467) = 2(-1.5467)^2 + 5(-1.5467) + 1$$
$$f(-1.5467) = 4.7846 - 7.7335 + 1$$
$$f(-4.2174) = -1.9489$$

Using the secant formula and substituting:

$$x_6 = -1.5467 - \frac{(-1.9489)[-1.5467 - (-4.2174)]}{-1.9489 - 15.4859}$$

$$x_6 = -1.5467 - \frac{(-5.2049)}{-17.4348} = -1.8452, este\ valor\ es\ el\ nuevo\ x_i$$

Determining the Percentage Error (Ep) with the formula No. 2, it is obtained:

$$E_p = \left|\frac{-1.8452 - (-1.5467)}{-1.8452}\right| * 100 = 16.18\%$$

A better percentage error than the previous ones, right?

Iteration No. 7. With the value of x_4 = -4.2174, the next iteration is determined considering that: $x_5 = -1.5467,\ x_6 = -1.8452$

$$x_7 = x_6 - \frac{f(x_6)(x_6 - x_5)}{f(x_6) - f(x_5)}$$

Evaluating f(x) we have:

$$f(x_5) = f(-1.5467) = 2(-1.5467)^2 + 5(-1.5467) + 1$$
$$f(-1.5467) = 4.7846 - 7.7335 + 1$$
$$f(-1.5467) = -1.9489$$

$$f(x_6) = f(-1.8452) = 2(-1.8452)^2 + 5(-1.8452) + 1$$
$$f(-1.8452) = 6.8095 - 9.226 + 1$$
$$f(-1.8452) = -1.4165$$

Using the secant formula and substituting:

$$x_7 = -1.8452 - \frac{(-1.4165)[-1.8452 - (-1.5467)]}{-1.4165 - (-1.9489)}$$

$$x_7 = -1.8452 - \frac{0.4228}{0.5324} = -2.6393, este\ valor\ es\ el\ nuevo\ x_i$$

Determining the Percentage Error (Ep) with the formula No. 2, it is obtained:

$$E_p = \left|\frac{-2.6393 - (-1.8452)}{-2.6393}\right| * 100 = 30.09\%$$

Have some patience and security in knowing that all the iterations done up to this point have been correct, even if he tries to prove otherwise. E_p tries to prove otherwise.

Iteration No. 8. With the value of x_4 = -4.2174, the next iteration is determined considering that: $x_6 = -1.8452,\ x_7 = -2.6393$

$$x_8 = x_7 - \frac{f(x_7)(x_7 - x_6)}{f(x_7) - f(x_6)}$$

Evaluating f(x) we have:

$$f(x_6) = f(-1.8452) = 2(-1.8452)^2 + 5(-1.8452) + 1$$
$$f(-1.8452) = 6.8095 - 9.226 + 1$$
$$f(-1.8452) = -1.4165$$

$$f(x_7) = f(-2.6393) = 2(-2.6393)^2 + 5(-2.6393) + 1$$
$$f(-2.6393) = 13.9318 - 13.1965 + 1$$
$$f(-2.6393) = 1.7353$$

Using the secant formula and substituting:

$$x_8 = -2.6393 - \frac{(1.7353)[-2.6393 - (-1.8452)]}{1.7353 - (-1.4165)}$$

$$x_8 = -2.6393 - \frac{(-1.3780)}{3.1518} = -2.2021, este\ valor\ es\ el\ nuevo\ x_i$$

Determining the Percentage Error (Ep) with the formula No. 2, it is obtained:

$$E_p = \left|\frac{-2.2021 - (-2.6393)}{-2.2021}\right| * 100 = 19.85\%$$

With the value obtained in iteration No. 8, where x_8 = -2.2021 a result in the percentage error again approaching 0% is observed.

Iteration No. 9. With the value of x_8 = -2.2021, the next iteration is determined considering that: $x_7 = -2.6393,\ x_8 = -2.2021$

$$x_9 = x_8 - \frac{f(x_8)(x_8 - x_7)}{f(x_8) - f(x_7)}$$

Evaluating f(x) we have:

$$f(x_7) = f(-2.6393) = 2(-2.6393)^2 + 5(-2.6393) + 1$$
$$f(-2.6393) = 13.9318 - 13.1965 + 1$$
$$f(-2.6393) = 1.7353$$

$$f(x_8) = f(-2.2021) = 2(-2.2021)^2 + 5(-2.2021) + 1$$
$$f(-2.2021) = 9.6985 - 11.0105 + 1$$
$$f(-2.2021) = -0.312$$

Using the secant formula and substituting:

$$x_9 = -2.2021 - \frac{(-0.312)[-2.2021 - (-2.6393)]}{-0.312 - (1.7353)}$$

$$x_9 = -2.2021 - \frac{(-0.1364)}{-2.0473} = -2.2687, este\ valor\ es\ el\ nuevo\ x_i$$

Determining the Percentage Error (Ep) with the formula No. 2, it is obtained:

$$E_p = \left|\frac{-2.2687 - (-2.2021)}{-2.2687}\right| * 100 = 2.94\%$$

How about the E_p? Much better, isn't it?

Iteration No. 10. With the value of x_9 = -2.2687, the next iteration is determined considering that: $x_8 = -2.2021,\ x_9 = -2.2687$

$$x_{10} = x_9 - \frac{f(x_9)(x_9 - x_8)}{f(x_9) - f(x_8)}$$

Evaluating f(x) we have:

$$f(x_8) = f(-2.2021) = 2(-2.2021)^2 + 5(-2.2021) + 1$$
$$f(-2.2021) = 9.6985 - 11.0105 + 1$$
$$f(-2.2021) = -0.312$$

$$f(x_9) = f(-2.2687) = 2(-2.2687)^2 + 5(-2.2687) + 1$$
$$f(-2.2687) = 10.2940 - 11.3435 + 1$$
$$f(-2.2687) = -0.0495$$

Using the secant formula and substituting:

$$x_{10} = -2.2687 - \frac{(-0.0495)[-2.2687 - (-2.2021)]}{-0.0495 - (-0.312)}$$

$$x_{10} = -2.2687 - \frac{0.00329}{0.2625} = -2.2812, este\ valor\ es\ el\ nuevo\ x_i$$

Determining the Percentage Error (Ep) with the formula No. 2, it is obtained:

$$E_p = \left|\frac{-2.2812 - (-2.2687)}{-2.2812}\right| * 100 = 0.55\%$$

We are very close to the closest approximation of the real value. We are already below 0%. If we continue with the iteration it is only to be as close as possible.

Iteration No. 11. With the value of x_{10} = -2.2812, the next iteration is determined considering that: $x_9 = -2.2687,\ x_{10} = -2.2812$

$$x_{11} = x_{10} - \frac{f(x_{10})(x_{10} - x_9)}{f(x_{10}) - f(x_9)}$$

Evaluating f(x) we have:

$$f(x_9) = f(-2.2687) = 2(-2.2687)^2 + 5(-2.2687) + 1$$
$$f(-2.2687) = 10.2940 - 11.3435 + 1$$
$$f(-2.2687) = -0.0495$$

$$f(x_{10}) = f(-2.2812) = 2(-2.2812)^2 + 5(-2.2812) + 1$$
$$f(-2.2812) = 10.4077 - 11.406 + 1$$
$$f(-2.2812) = 0.0017$$

Using the secant formula and substituting:

$$x_{11} = -2.2812 - \frac{(0.0017)[-2.2812 - (-2.2687)]}{0.0017 - (-0.0495)}$$

$$x_{11} = -2.2812 - \frac{(-0.00002125)}{0.0512}$$
$$= -2.2808, este\ valor\ es\ el\ nuevo\ x_i$$

Determining the Percentage Error (Ep) with the formula No. 2, it is obtained:

$$E_p = \left|\frac{-2.2808 - (-2.2812)}{-2.2808}\right| * 100 = 0.018\%$$

There, we have arrived; although it is not a pure 0% but it is too close, this will be demonstrated by the following check.

- **Check.**

  Substitute the value x_{11} = -2.2808 in the function:

  f(x) = $2x^2 + 5x + 1$

  $2(-2.2808)^2 + 5(-2.2808) + 1 = 0$

$$10.404 - 11.404 + 1 = 0$$

$$0.000 = 0$$

The value $\mathbf{x_{11} = -2.2808}$ obtained in iteration 11 can then already be considered correct. Congratulations!

The following excel table 5 shows all the possible iterations for this function, as well as the variation and approximation obtained on the Ep.

Método de la Secante de Iteraciones

Iteración	Xo	Xn	X1	f(xo)	f(x1)	Ea	Er	Ep
1	-2	-1.66667	1	-1	8			
2	1	-1.18182	-1.66667	8	-1.77778	0.48485	0.41026	41.03%
3	-1.66667	-4.21739	-1.18182	-1.77778	-2.11570	-3.03557	0.71978	71.98%
4	-1.18182	-1.54669	-4.21739	-2.11570	15.48582	2.67070	1.72671	172.67%
5	-4.21739	-1.84524	-1.54669	15.48582	-1.94895	-0.29854	0.16179	16.18%
6	-1.54669	-2.63924	-1.84524	-1.94895	-1.41638	-0.79400	0.30084	30.08%
7	-1.84524	-2.20210	-2.63924	-1.41638	1.73495	0.43713	0.19851	19.85%
8	-2.63924	-2.26873	-2.20210	1.73495	-0.31200	-0.06663	0.02937	2.94%
9	-2.20210	-2.28126	-2.26873	-0.31200	-0.04937	-0.01253	0.00549	0.55%
10	-2.26873	-2.28077	-2.28126	-0.04937	0.00198	0.00048	0.00021	0.02%
11	-2.28126	-2.28078	-2.28077	0.00198	-0.00001	0.00000	0.00000	0.00%
12	-2.28077	-2.28078	-2.28078	-0.00001	0.00000	0.00000	0.00000	0.00%
13	-2.28078	-2.28078	-2.28078	0.00000	0.00000	0.00000	0.00000	0.00%
14	-2.28078	-2.28078	-2.28078	0.00000	0.00000	0.00000	0.00000	0.00%
15	-2.28078	-2.28078	-2.28078	0.00000	0.00000	0.00000	0.00000	0.00%

... Falsa Posición | Punto Fijo | Newton Raphson | Secante

Table 5. Excel sheet - Secant method

Table 5 shows that from iteration number 11 onwards, the root value calculations remain stable; from iteration 11 to 15, the quantity -2.28078 is still maintained, thus confirming the value ascertained in the previous paragraphs.

CONCLUSIONS OF THE METHOD:

Probably because it has taken us to eleven iterations this method loses its appeal for many, largely due to the interval range that was chosen; if you have the opportunity to automate this function in an excel sheet you could vary the initial values and you will notice that in some cases fewer intervals are required.

The secant method is not complex, as you could observe, it was only necessary to evaluate the initial values in the function and then substitute it in the formula number 1 of the secant, this last one could be the most complicated if we do not know how to substitute values.

At the end of it all we were able to come up with the value that satisfies the function, I hope you find it useful.

2.4 MULTIPLE ROOT METHOD

The Multiple Roots Method is different compared to all the previous methods that have been described, the substantive part is centred on the factorisation of a function and on analysing the number of factors that are repeated, since the interpretation that can and should be given to the graph of f(x) depends on this; unlike the previous methods, the Multiple Roots Method will always provide the point of tangency that the abscissa axis generates with f(x).

A multiple root corresponds to a point where a function is tangent to the **X-axis** (Chapra & P. Canale, 2007)For example, a double root (multiple root) is as shown in the following expression:

$$f(x) = (x-3)(x-1)(x-1)$$

From this expression, three factors can be observed, where x - 1 are repeated twice. The factors $(x-3)(x-1)(x-1)$ are the result of factoring the expression representing the function: $f(x) = x^3 - 5x^2 + 7x - 3$

If you wanted to check that $(x-3)(x-1)(x-1) = x^3 - 5x^2 + 7x - 3$ they are equal, just multiply $= (x-3)(x-1)^2$ where $(x-1)^2 = (x-1)(x-1)$

Multiplying these factors, it can be seen that:

$$f(x) = (x-3)(x^2 - 2x + 1)$$

$$f(x) = x^3 - 5x^2 + 7x - 3$$

But what is the use of knowing the number of roots or equal factors that a polynomial has, the answer is interesting, because those repeated roots indicate the point at which the X axis of the Cartesian plane is tangent with the graph of the function, that is to say, analytically we can conclude that the *method of multiple roots* at no time will help us to determine at what point the graph intersects the X axis as the previously studied methods allow us to do; In other words, it can indicate a maximum or minimum of the function where the reference will always be the abscissa axis or, if necessary, a possible inflection point.

Let's analyse this graphically, to demonstrate what the previous paragraph intends to share we will use the classic Geogebra application version 5.0 to graph the function f(x) above:

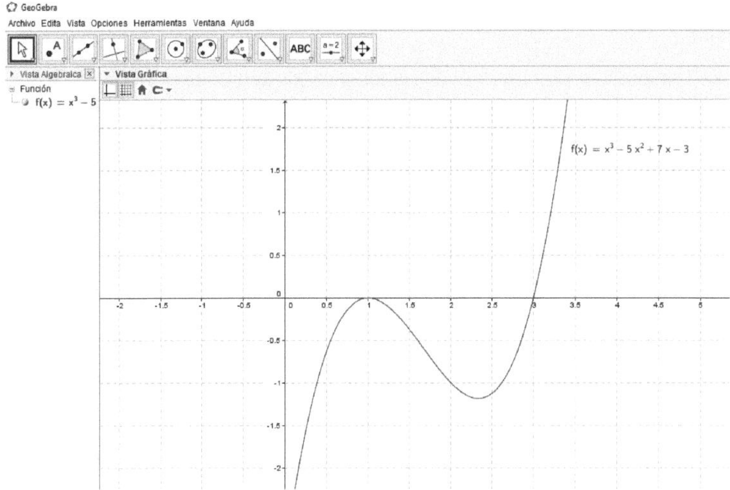

Figure No. 2 Graph of the function f(x)

Note that the graph of the function has a tangent line at the point (1,0), but coincidentally this line is the abscissa axis of the Cartesian plane, this is referred to in the *Multiple Root Method* for obtaining roots of an equation.

To be clear, from the function $f(x) = (x-3)(x-1)(x-1)$The factors (x - 1) which are repeated twice, indicate to me the point where there will be a tangent line to the curve of the function f(x) and that this line will be the abscissa axis of the Cartesian plane and this can be demonstrated by equalling zero and clearing the variable X, as described below:

$$x - 1 = 0$$

$$x = 1$$

Then, the tangent line to the graph of the function f(x) is given at x = 1 and y = 0, so the point is (1, 0) as shown in the graph.

Reciprocally, the intersection point (3, 0) that the graph gives, is obtained from the factor (x - 3), after equating to 0 we obtain that x = 3 and y = 0.

Any other method previously studied, such as Newton Raphson, secant, etc., would be somewhat inefficient and incapable of determining the point where the tangency occurs, due to the characteristics of the function itself, likewise, these methods lie their strength in determining intersections with the abscissa and not a change of direction of the graph as seen in the graph that geogebra provides us with.

Looked at another way, at (1,0) is the local maximum of the function f(x).

Let us look at other examples:

Sea $f(x) = x^4 - 4x^2 + 4$ (Olvera, 2001)

When this function is decomposed into factors, it is concluded that:

$$f(x) = (x^2 - 2)^2 = (x^2 - 2)(x^2 - 2)$$

Note that each factor is a difference of squares, so that when decomposed:

$$f(x) = (x - \sqrt{2})(x + \sqrt{2})(x - \sqrt{2})(x + \sqrt{2})$$

It may be noted that there are two pairs of repeating factors; as multiple roots, they then indicate that each pair constructs a point on the graph of the function where the abscissa axis is tangent to the graph of the function f(x).

$$(x - \sqrt{2})(x - \sqrt{2})$$

Of these two, only one of the factors is considered, when equalled to zero, it is concluded that:

$$x - \sqrt{2} = 0$$

$x = \sqrt{2}$ therefore, the point of tangency is given at $(\sqrt{2}, 0)$, reciprocally, the other pair:

$$(x + \sqrt{2})(x + \sqrt{2})$$

$$x + \sqrt{2} = 0$$
$$x = -\sqrt{2} \therefore (-\sqrt{2}, 0)$$

Let's look at the graph of the function and analyse these results:

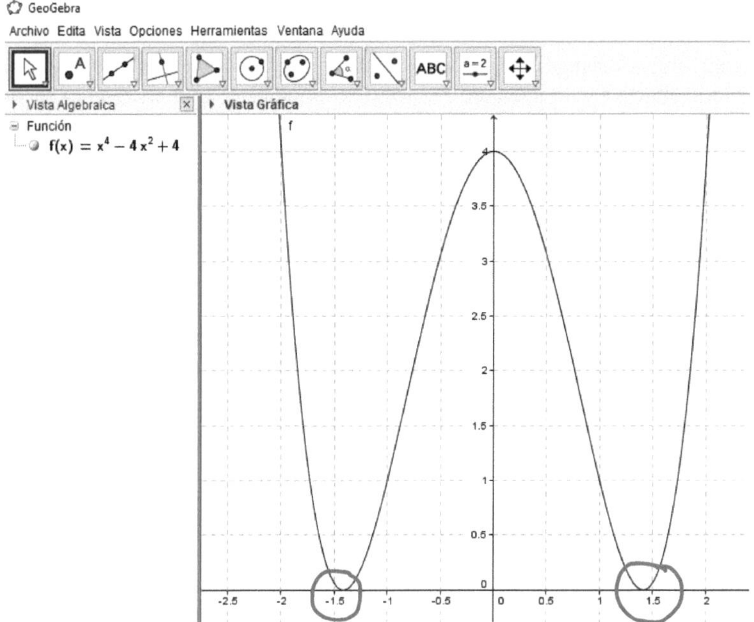

Figure No. 3 Graph of x^4 - $4x^2$ + 4

The parts indicated in the image represent the multiple roots of the function f(x), it can be clearly seen that the x-axis of the Cartesian plane is tangent to the curves of the function, looking at it another way, two relative minima are constructed, although the latter would not necessarily have to be so, also, as indicated in the graph that geogebra provides us, they are equivalent to: $(-\sqrt{2}, 0)$ y $(\sqrt{2}, 0)$

CONCLUSIONS

In this book, we have explored the most relevant numerical methods for solving complex equations and functions, intentionally solving the same function f(x) in order to demonstrate that all methods lead to the same result. Each method has its advantages and disadvantages, and the choice of the most appropriate one will depend on the type of function that needs to be solved.

It is essential to emphasise that mastery of basic mathematical skills, such as algebra and analytical geometry, is essential to understand and apply these methods effectively. In addition, the use of technological tools such as Excel and GeoGebra 5.0 can significantly facilitate the problem-solving process.

In summary, Numerical Methods are powerful tools for solving complex problems in various areas of science and engineering. We hope that this book has provided a solid foundation for their understanding and application, and that readers are now better equipped to meet the mathematical challenges they face.

Bibliography

Chapra, S. C., & P. Canale, R. (2007). *Numerical methods for engineers* (5th ed., Vol. Unique). Mexico, DF, Mexico: Mc Graw Hill.

Nieves, A., & C. Domínguez, F. (1998). *Numerical Methods Applied to Engineering.* Mexico, DF: CECSA.

Olvera, B. G. (2001). *Mathematics - Differential Calculus.* Mexico, DF: DGTI.

Printed by Books on Demand GmbH, Norderstedt / Germany